YOUR KNOWLEDGE HAS VALUE

AF125073

- We will publish your bachelor's and master's thesis, essays and papers

- Your own eBook and book - sold worldwide in all relevant shops

- Earn money with each sale

Upload your text at www.GRIN.com and publish for free

Bibliographic information published by the German National Library:

The German National Library lists this publication in the National Bibliography; detailed bibliographic data are available on the Internet at http://dnb.dnb.de .

This book is copyright material and must not be copied, reproduced, transferred, distributed, leased, licensed or publicly performed or used in any way except as specifically permitted in writing by the publishers, as allowed under the terms and conditions under which it was purchased or as strictly permitted by applicable copyright law. Any unauthorized distribution or use of this text may be a direct infringement of the author s and publisher s rights and those responsible may be liable in law accordingly.

Imprint:

Copyright © 2019 GRIN Verlag
Print and binding: Books on Demand GmbH, Norderstedt Germany
ISBN: 9783668932937

This book at GRIN:

https://www.grin.com/document/464786

Deapon Biswas

Summation Methods

GRIN Verlag

GRIN - Your knowledge has value

Since its foundation in 1998, GRIN has specialized in publishing academic texts by students, college teachers and other academics as e-book and printed book. The website www.grin.com is an ideal platform for presenting term papers, final papers, scientific essays, dissertations and specialist books.

Visit us on the internet:

http://www.grin.com/

http://www.facebook.com/grincom

http://www.twitter.com/grin_com

The Summation Methods

Deapon Biswas
Transport Officer, Private Concern, Chattogram, Bangladesh.

Abstract

In Mathematics interval is a range of numbers between two given numbers and indexes are the including numbers between those two numbers. This paper discusses the beautiful and effective applications of indexes and intervals. I discuss this topic based on the theory of summation methods. Here I tried to show how addition and multiplication are closely connected.

Keywords

Index, interval, possible events and selected events, summation methods, Biswas triangle.

Article outline

1. Introduction
2. Index
3. Summation method I
4. Summation method II
5. Summation method III
6. Biswas triangle
7. Summation method IV
8. Conclusions

1. Introduction

Here I define all necessary terms to complete the paper. And then develop 5 theorems of which 4 are for summation methods and 1 is for Biswas triangle. Biswas triangle is a new concept introduced in this paper.

2. Index

Let a parent assembly
$$A = (A_1, A_2, A_3, ..., A_i, ..., A_R)$$
and an infant assembly
$$B_i = (B_{i1}, B_{i2}, B_{i3}, ..., B_{iv}, ..., B_{iV})$$

then k_v is an index that indicates a component of the parent assembly A taken by v^{th} place of the infant assembly B_i.

Example 1: Let a parent assembly A = (a, b, c, d, e) and the B members (i) B_i = (a, c, e), (ii) AB_i = (e, a, c). Find the indexes with their indicated components.

Solution: (i) For B_i $k_1 = 1$ indicates a

                                $k_2 = 3$ indicates c

                                $k_3 = 5$ indicates e

        (ii) For AB_i $k_1 = 5$ indicates e

                                $k_2 = 1$ indicates a

                                $k_3 = 3$ indicates c

2.1 Effective intervals: Let the interval that indicates only integers

$$1 \le k_v \le R \qquad\qquad\qquad\qquad (1)$$

where k_v varies from 1 to R. Then the interval is said to be effective interval. Again the intervals that indicate only integers

$$1 \le k_v \le k_1 - 1$$
$$k_1 + 1 \le k_v \le k_2 - 1$$
$$k_2 + 1 \le k_v \le k_3 - 1 \qquad\qquad\qquad (2)$$
$$\vdots$$
$$k_{v-1} + 1 \le k_v \le R$$

where the condition $1 < k_1 < k_2 < k_3 < ... < k_{v-1} < R$ supports then the intervals are to be said effective intervals. The assembly containing the effective intervals denoted by E_v. Thus we can say there are two types of effective intervals, one is simple effective interval and other is conditional effective interval. The assembly of simple effective interval E_v contains one interval and the assembly of conditional effective intervals E_v contains many many intervals. The first one relates in the discussion for combinations and the second one relates in the discussion for permutations. The intervals are said to be on possible forms.

Example 2: Find the effective intervals in an assembly of effective intervals E_4.

Solution: The simple effective interval in the given assembly is

$$1 \le k_4 \le R$$

and the conditional effective intervals in the given assembly are

$1 \leq k_4 \leq k_1-1$

$k_1+1 \leq k_4 \leq k_2-1$

$k_2+1 \leq k_4 \leq k_3-1$

$k_3+1 \leq k_4 \leq R$

subject to the condition $1 < k_1 < k_2 < k_3 < R$.

2.2 Possible intervals: Let the effective intervals

$1 \leq k_v \leq k_1-1$

$k_1+1 \leq k_v \leq k_2-1$

$k_2+1 \leq k_v \leq k_3-1$

$\vdots$

$k_{v-1}+1 \leq k_v \leq R$

where the condition $1 < k_1 < k_2 < k_3 < ... < k_{v-1} < R$ supports. If the condition varies then the form of effective intervals are also varies. These forms are to be called possible intervals. The assembly containing the possible intervals denoted by F_v. Clearly there are two types of possible intervals, one is simple possible interval and another is conditional possible interval. The simple effective interval makes the possible interval itself thus the assembly F_v contains one interval. The conditional effective intervals make many many possible intervals due to the condition varies. Thus the assembly F_v contains many many intervals. The intervals are said to be on possible forms and found from effective intervals.

Example 3: Find the possible intervals in an assembly of possible intervals F_4.

Solution: The simple possible interval in the given assembly is

$1 \leq k_4 \leq R$

and the conditional possible intervals in the assembly are

$1 \leq k_4 \leq k_1-1, \ 1 \leq k_4 \leq k_2-1, \ 1 \leq k_4 \leq k_3-1, \ k_1+1 \leq k_4 \leq k_2-1,$

$k_1+1 \leq k_4 \leq k_3-1, \ k_1+1 \leq k_4 \leq R, \ k_2+1 \leq k_4 \leq k_1-1, \ k_2+1 \leq k_4 \leq k_3-1,$

$k_2+1 \leq k_4 \leq R, \ k_3+1 \leq k_4 \leq k_1-1, \ k_3+1 \leq k_4 \leq k_2-1, \ k_3+1 \leq k_4 \leq R.$

2.3 Particular intervals: When a possible interval takes particular values for lower and upper limits then the interval is said to be particular interval.

The assembly containing particular intervals denoted by W_v. For investigation there contains in the assembly W_v alike intervals and they should be counted. As the particular intervals are made from possible intervals so there are two types of particular intervals, one is simple particular interval and other is conditional particular interval. The intervals are said to be on possible forms and found from possible intervals.

Example 4: Find the particular intervals in an assembly of particular intervals W_4 where $R = 5$, $k_1 = 1$, $k_2 = 2$ and $k_3 = 3$.
Solution: The simple particular interval in the given assembly is
$$1 \leq k_4 \leq 5$$
and the conditional particular intervals in the given assembly are
$$1 \leq k_4 \leq 1,\ 1 \leq k_4 \leq 2,\ 2 \leq k_4 \leq 2,\ 2 \leq k_4 \leq 5,\ 3 \leq k_4 \leq 5,\ 4 \leq k_4 \leq 5.$$
The above conditional particular intervals are found from example 3 putting the values of R, k_1, k_2 and k_3. Here vanishes the intervals when upper limit found less than lower limit.

2.4 Assembly of assemblies of indexes: We know the intervals contain only indexes (integers) thus the assembly of indexes is the assembly that contains indexes of an interval. It is denoted by $K_{v.g}$ i.e.,
$$K_{v.g} = (1, 2, 3, \ldots., R) \hspace{2cm} \text{————————— (3)}$$
The assembly is said to be on possible form. Now the assembly of assemblies of indexes is the assembly that contains the assemblies of indexes $K_{v.1}$, $K_{v.2}$, $K_{v.3}$, ….., $K_{v.g}$, ..., $K_{v.h}$. The assembly is denoted by K_v i.e.,
$$K_v = (K_{v.1}, K_{v.2}, K_{v.3}, \ldots., K_{v.g}, ..., K_{v.h}) \hspace{1cm} \text{————————— (4)}$$
The assembly is said to be on possible form. Like intervals there are two types of assemblies of assemblies of indexes, one is simple assembly of assemblies of indexes and other is conditional assembly of assemblies of indexes. These assemblies are found from particular intervals.

Example 5: Find the assembly of assemblies of indexes K_4 from the example 4.
Solution: The simple assembly of assemblies of indexes is
$$K_4 = ((4, 5))$$
and the conditional assembly of assemblies of indexes is

$K_4 = ((1), (1, 2), (2), (2, 3, 4, 5), (3, 4, 5), (4, 5))$.

2.5 Selected forms of effective intervals: We have discussed the possible forms of intervals. Now we shall discuss the selected forms of intervals. Let the simple effective interval

$1 \leq k_v \leq R$ [see (1)]

or the conditional effective intervals

$1 \leq k_v \leq k_1 - 1$

$k_1 + 1 \leq k_v \leq k_2 - 1$

$k_2 + 1 \leq k_v \leq k_3 - 1$ [see (2)]

$\vdots$

$k_{v-1} + 1 \leq k_v \leq R$

where the intervals contain every integers from lower limit to upper limit. When the intervals contain some specific integers by means and some are not from the intervals (1) and (2) then the intervals containing specific integers are to be called selected forms of effective intervals. Like possible forms there are two types of effective intervals, one is simple effective interval on selected forms and other is conditional effective interval on selected forms. The first type expressed by

$1 \leq k_v^* \leq \text{till } R$ ————————— (5)

and the second type expressed by

$1 \leq k_v^* \leq \text{till } k_1^* - 1$

$k_1^* + 1 \leq k_v^* \leq \text{till } k_2^* - 1$

$k_2^* + 1 \leq k_v^* \leq \text{till } k_3^* - 1$ ————————— (6)

$\vdots$

$k_{v-1}^* + 1 \leq k_v^* \leq \text{till } R$

subject to the condition $1 < k_1^* < k_2^* < k_3^* < ... < k_{v-1}^* < R$. The assembly containing the effective intervals on selected form is denoted by E_v^*.

Example 6: Find the assembly of effective intervals on selected forms E_4^*.

Solution: The assembly of simple effective interval on selected forms is

$E_4^* = (1 \leq k_4^* \leq \text{till } R)$

and the assembly of conditional effective intervals on selected forms is

$E_4^* = (1 \leq k_4^* \leq$ till $k_1^* - 1$, $k_1^* + 1 \leq k_4^* \leq$ till $k_2^* - 1$, $k_2^* + 1 \leq k_4^* \leq$ till $k_3^* - 1$, $k_3^* + 1 \leq k_4^* \leq$ till R)

subject to the condition $1 < k_1^* < k_2^* < k_3^* < R$.

2.6 Selected forms of possible intervals: Suppose the effective intervals (6) where the condition is $1 < k_1^* < k_2^* < k_3^* < ... < k_{v-1}^* < R$. If the condition varies as $1 < k_3^* < k_2^* < k_1^* < ... < k_{v-1}^* < R$ then the effective intervals on selected forms can be written as

$1 \leq k_v^* \leq$ till $k_3^* - 1$

$k_3^* + 1 \leq k_v^* \leq$ till $k_2^* - 1$

$k_2^* + 1 \leq k_v^* \leq$ till $k_1^* - 1$

$\vdots$

$k_{v-1}^* + 1 \leq k_v^* \leq$ till R

Thus we see there are many many assemblies of effective intervals on selected forms if the condition demands. These effective intervals in totally are to be called possible intervals on selected forms or selected forms of possible intervals. Like possible forms there are two types of possible intervals, one is simple possible interval on selected forms and other is conditional possible interval on selected forms. The intervals are found from effective intervals. The assembly containing possible intervals on selected forms is denoted by F_v^*.

Example 7: Find the assembly of possible intervals on selected form F_4^*.
Solution: The assembly of simple possible interval on selected forms is
$F_4^* = (1 \leq k_4^* \leq$ till R)
and the assembly of conditional possible intervals on selected forms is
$F_4^* = (1 \leq k_4^* \leq$ till $k_1^* - 1$, $1 \leq k_4^* \leq$ till $k_2^* - 1$, $1 \leq k_4^* \leq$ till $k_3^* - 1$, $k_1^* + 1 < k_4^* \leq$ till $k_2^* - 1$, $k_1^* + 1 \leq k_4^* \leq$ till $k_3^* - 1$, $k_1^* + 1 \leq k_4^* \leq$ till R, $k_2^* + 1 \leq k_4^* \leq$ till $k_1^* - 1$, $k_2^* + 1 \leq k_4^* \leq$ till $k_3^* - 1$, $k_2^* + 1 \leq k_4^* \leq$ till R, $k_3^* + 1 \leq k_4^* \leq$ till $k_1^* - 1$, $k_3^* + 1 \leq k_4^* \leq$ till $k_2^* - 1$, $k_3^* + 1 \leq k_4^* \leq$ till R).

2.7 Selected form of particular intervals: When a possible interval on selected form takes particular values for lower and upper limits then the interval is said to be particular interval on selected form or selected form of particular interval. Like possible forms there are two types of particular intervals, one is simple particular interval on selected forms and other is conditional particular interval on selected forms. The intervals are found from possible intervals. The assembly containing particular intervals on selected forms is denoted by W_v^*. When the upper limit found less than lower limit then the interval vanishes.

Example 8: Find the assembly of particular intervals on selected forms W_4^* where R= 10, $k_1^* = 3$, $k_2^* = 7$ and $k_3^* = 5$.

Solution: The assembly of simple particular intervals on selected form is

$W_4^* = (1 \leq k_4^* \leq \text{till } 10)$

Again the assembly of conditional particular intervals on selected form is

$W_4^* = (1 \leq k_4^* \leq \text{till } 2, 1 \leq k_4^* \leq \text{till } 6, 1 \leq k_4^* \leq \text{till } 4, 4 \leq k_4^* \leq \text{till } 6, 4 \leq k_4^*$ $\leq \text{till } 4, 4 \leq k_4^* \leq \text{till } 10, 8 \leq k_4^* \leq \text{till } 10, 6 \leq k_4^* \leq \text{till } 6, 6 \leq k_4^* \leq \text{till } 10)$

where the condition is $k_1^* < k_3^* < k_2^* < R$.

The solution is found from example 7 and the intervals when upper limit found less than lower limit vanish.

2.8 Possible indexes and selected indexes: Let the effective intervals on possible forms

$1 \leq k_v \leq R$

or,

$1 \leq k_v \leq k_1 - 1$

$k_1 + 1 \leq k_v \leq k_2 - 1$

$k_2 + 1 \leq k_v \leq k_3 - 1$

$\vdots$

$k_{v-1} + 1 \leq k_v \leq R$

subject to the condition $1 < k_1 < k_2 < k_3 < ... < k_{v-1} < R$. Then the indexes contained in these intervals are to be called possible indexes. The indexes are denoted by k_v. Again the effective intervals on selected forms

$1 \leq k_v^* \leq \text{till } R$

or,

$1 \leq k_v^* \leq \text{till } k_1^* - 1$

$k_1^* + 1 \leq k_v^* \leq \text{till } k_2^* - 1$

$k_2^* + 1 \leq k_v^* \leq \text{till } k_3^* - 1$

$\vdots$

$k_{v-1}^* + 1 \leq k_v^* \leq$ till R

subject to the condition $1 < k_1^* < k_2^* < k_3^* < ... < k_{v-1}^* < R$. Then the indexes contained in these intervals are to be called selected indexes. The indexes are denoted by k_v^*. Remember the selected indexes are some specific integers by means and the lower limit must be contained as a selected index.

Example 9: Let an assembly A = (a, a, b, b, b, c, d, d). Find the possible indexes contained in

$1 \leq k_3 \leq R$ and contained in $1 \leq k_3 \leq k_1 - 1$, $k_1 + 1 \leq k_3 \leq k_2 - 1$, $k_2 + 1 \leq k_3$ $\leq R$. Again find the selected indexes contained in these intervals where $k_1 = 1$ and $k_2 = 3$ in which selected indexes start a kind in the assembly A except the lower limits.

Solution: The possible indexes are given after semicolon's of their respective intervals

$1 \leq k_3 \leq R$	;	1, 2, 3, 4, 5, 6, 7, 8
$1 \leq k_3 \leq k_1 - 1$	;	——————
$k_1 + 1 \leq k_3 \leq k_2 - 1$	;	2
$k_2 + 1 \leq k_3 \leq R$	;	4, 5, 6, 7, 8

The selected indexes are given after semicolon's of their respective intervals

$1 \leq k_3^* \leq$ till R	;	1, 3, 6, 7
$1 \leq k_3^* \leq$ till $k_1^* - 1$	;	———
$k_1^* + 1 \leq k_3^* \leq$ till $k_2^* - 1$	;	2
$k_2^* + 1 \leq k_3^* \leq$ till R	;	4, 6, 7

2.9 Assembly of assemblies of selected indexes: The assembly of selected indexes is the assembly that contains selected indexes of an interval on selected forms. It is denoted by $k_{v.g}^*$ i.e.,

$$k_{v.g}^* = (1, R_1 + 1, R_1 + R_2 + 1, ..., R_1 + R_2 + R_3 + ... + R_{h-1} + 1) \quad \text{———— (7)}$$

Now the assembly of assemblies of selected indexes is the assembly that contains the assemblies of selected indexes $k_{v.1}^*$, $k_{v.2}^*$, $k_{v.3}^*$, ..., $k_{v.g}^*$, ..., $k_{v.h}^*$. The assembly is denoted by K_v^* i.e.,

$$K_v^* = (K_{v.1}^*, K_{v.2}^*, K_{v.3}^*, ..., K_{v.g}^*, ..., K_{v.h}^*) \quad \text{——————— (8)}$$

Like intervals there are two types of assemblies of assemblies of selected indexes, one is simple assembly of assemblies of selected indexes and other is conditional assembly of assemblies of selected indexes. These assemblies are found from particular intervals on selected forms.

Example 10: Find the assembly of assemblies of selected indexes K_3^* from the example 6.

Solution: The simple assembly of assemblies of selected indexes is

$K_3^* = ((1, 3, 6, 7))$

and the conditional assembly of assemblies of selected indexes is

$K_3^* = ((2), (4, 6, 7))$.

2.10 Possible events and selected events: Let a parent assembly that of different components then the events characterizing v^{th} components are to be all different. These events are called possible events. The number of possible events is denoted by s_v. Again let a parent assembly that the components not all different then there occurred some alike events in the event space $B\left\{{A \atop V/v}\right\}$. The different events to one another are to be called selected events. The number of selected events is denoted by s_v^*.

Example 11: Let the parent assembly $A = (a, a, b, b, b, c, d, d)$. Now find the possible events and selected events of A where the members are to be combinations of 6 components characterizing third components.

Solution: The possible events are

(i) $B\left\{{A \atop 6/(a, a, b)}\right\}$ = {(a, a, b, b, b, c), (a, a, b, b, b, d), (a, a, b, b, b, d), (a, a, b, b, c, d), (a, a, b, b, c, d), (a, a, b, b, d, d), (a, a, b, b, c, d), (a, a, b, b, c, d), (a, a, b, b, d, d), (a, a, b, c, d, d)}

(ii) $B\left\{{A \atop 6/(a, a, b)}\right\}$ = {(a, a, b, b, c, d), (a, a, b, b, c, d), (a, a, b, b, c, d), (a, a, b, c, d, d)}

(iii) $B\left\{{A \atop 6/(a, a, b)}\right\}$ = {(a, a, b, c, d, d)}

(iv) $B\left\{{A \atop 6/(a, b, b)}\right\}$ = {(a, b, b, b, c, d), (a, b, b, b, c, d), (a, b, b, b, d, d), (a, b, b, c, d, d)}

(v) $B\left\{{A \atop 6/(a, b, b)}\right\}$ = {(a, b, b, c, d, d)}

(vi) $B\left\{\begin{matrix} A \\ 6/(a,b,b) \end{matrix}\right\} = \{(a, b, b, c, d, d)\}$

(vii) $B\left\{\begin{matrix} A \\ 6/(a,b,b) \end{matrix}\right\} = \{(a, b, b, b, c, d), (a, b, b, b, c, d), (a, b, b, b, d, d),$
$(a, b, b, c, d, d)\}$

(viii) $B\left\{\begin{matrix} A \\ 6/(a,b,b) \end{matrix}\right\} = \{(a, b, b, c, d, d)\}$

(ix) $B\left\{\begin{matrix} A \\ 6/(a,b,b) \end{matrix}\right\} = \{(a, b, b, c, d, d)\}$

(x) $B\left\{\begin{matrix} A \\ 6/(b,b,b) \end{matrix}\right\} = \{(b, b, b, c, d, d)\}$

The selected events are

(i) $B^*\left\{\begin{matrix} A \\ 6/(a,a,b) \end{matrix}\right\} = \{(a, a, b, b, b, c), (a, a, b, b, b, d), (a, a, b, b, c, d),$
$(a, a, b, b, d, d), (a, a, b, c, d, d)\}$

(ii) $B^*\left\{\begin{matrix} A \\ 6/(a,b,b) \end{matrix}\right\} = \{(a, b, b, b, c, d), (a, b, b, b, d, d), (a, b, b, c, d, d)\}$

(iii) $B^*\left\{\begin{matrix} A \\ 6/(b,b,b) \end{matrix}\right\} = \{(b, b, b, c, d, d)\}.$

3. Summation method I

The V-times summation of a constant quantity is equal to the product of that constant and the numbers of integers from lower limit to upper limit of the summation symbols, i.e.,

$$\sum_{k_1=l_1}^{h_1} \sum_{k_2=l_2}^{h_2} \sum_{k_3=l_3}^{h_3} \cdots \sum_{k_v=l_v}^{h_v} C = (h_1 - l_1 + 1) \times (h_2 - l_2 + 1) \times (h_3 - l_3 + 1)$$
$$\times \cdots \times (h_v - l_v + 1) \times C \quad \text{———— (9)}$$

where C is a constant quantity taking unit value.

Proof: We know

$$\sum_{k_1=l_1}^{h_1} x_{k_1} = x_{l_1} + x_{l_1+1} + x_{l_1+2} + \ldots + x_{h_1} \quad \text{———————— (10)}$$

Taking $x_{l_1} = x_{l_1+1} = x_{l_1+2} = \ldots = x_{h_1} = C$ then (10) becomes

$$\sum_{k_1=l_1}^{h_1} C = (h_1 - l_1 + 1) \times C$$

Again

$$\sum_{k_1=l_1}^{h_1} \sum_{k_2=l_2}^{h_2} x_{k_1 k_2} = \sum_{k_2=l_2}^{h_2} x_{l_1 k_2} + \sum_{k_2=l_2}^{h_2} x_{(l_1+1)k_2} + \sum_{k_2=l_2}^{h_2} x_{(l_1+2)k_2}$$

$$+.....+ \sum_{k_2=l_2}^{h_2} x_{h_1 k_2} \quad\quad\quad\quad\quad\quad (11)$$

Taking $x_{l_1 k_2} = x_{(l_1+1)k_2} = x_{(l_1+2)k_2} = ... = x_{h_1 k_2} = C$

then (11) becomes

$$\sum_{k_1=l_1}^{h_1} \sum_{k_2=l_2}^{h_2} C = \sum_{k_2=l_2}^{h_2} C + \sum_{k_2=l_2}^{h_2} C + \sum_{k_2=l_2}^{h_2} C +...+ \sum_{k_2=l_2}^{h_2} C$$

$$[(h_1 - l_1 + 1)\ \text{terms}]$$

$$= (h_1 - l_1 + 1)\sum_{k_2=l_2}^{h_2} C$$

$$= (h_1 - l_1 + 1) \times (h_2 - l_2 + 1) \times C$$

Similarly

$$\sum_{k_1=l_1}^{h_1} \sum_{k_2=l_2}^{h_2} \sum_{k_3=l_3}^{h_3} C = (h_1 - l_1 + 1) \times (h_2 - l_2 + 1) \times (h_3 - l_3 + 1) \times C$$

And so on by induction

$$\sum_{k_1=l_1}^{h_1} \sum_{k_2=l_2}^{h_2} \sum_{k_3=l_3}^{h_3} \cdots \sum_{k_v=l_v}^{h_v} C = (h_1 - l_1 + 1) \times (h_2 - l_2 + 1) \times (h_3 - l_3 + 1)$$

$$\times ...... \times (h_v - l_v + 1) \times C$$

Example 12: Prove that

$$\sum_{k_1=1}^{a_1} \sum_{k_2=1}^{a_2} \sum_{k_3=1}^{a_3} \cdots \sum_{k_v=1}^{a_v} C = a_1 \times a_2 \times a_3 \times ... \times a_v \times C$$

Solution: We know from (9)

$$\sum_{k_1=l_1}^{h_1} \sum_{k_2=l_2}^{h_2} \sum_{k_3=l_3}^{h_3} \cdots \sum_{k_v=l_v}^{h_v} C = (h_1 - l_1 + 1) \times (h_2 - l_2 + 1) \times (h_3 - l_3 + 1)$$

$$\times ... \times (h_v - l_v + 1) \times C \quad\quad (12)$$

Now taking $l_1 = 1, h_1 = a_1, l_2 = 1, h_2 = a_2, l_3 = 1, h_3 = a_3, ..., l_v = 1, h_v = a_v$

then (12) becomes

$$\sum_{k_1=1}^{a_1} \sum_{k_2=1}^{a_2} \sum_{k_3=1}^{a_3} \cdots \sum_{k_v=1}^{a_v} C = (a_1 - 1 + 1) \times (a_2 - 1 + 1) \times (a_3 - 1 + 1)$$

$$\times ... \times (a_v - 1 + 1) \times C$$

$$= a_1 \times a_2 \times a_3 \times ... \times a_v \times C$$

Example 13: Find $\sum_{k_1=1}^{3} \sum_{k_2=3}^{5} \sum_{k_3=3}^{4} C$

Solution: $\sum_{k_1=1}^{3} \sum_{k_2=3}^{5} \sum_{k_3=3}^{4} C = (3 - 1 + 1) \times (5 - 3 + 1) \times (4 - 3 + 1) \times C$

$$= 18C.$$

4. Summation method II

Let an operation can be performed in a_1 ways and after it is performed in any one of these ways a second operation can be performed in a_2 ways and after it is performed in any one of these ways a third operation can be performed in a_3 ways and so on for V operations, then the V operations can be performed together in

$$\sum_{k_1=1}^{a_1} \sum_{k_2=1}^{a_2} \sum_{k_3=1}^{a_3} \cdots \sum_{k_v=1}^{a_v} C \text{ ways} \qquad\qquad ——————— (13)$$

where C is a constant quantity taking unit value.

Proof: Let we have V operations to be performed. The first operation can be performed in to say a_1 ways. Then we can write the number of ways in which first operation performed is $\sum_{k_1=1}^{a_1} C$ where C is a constant quantity taking unit value. After it has been done the second operation can be performed in to say a_2 ways. Then we can write the number of ways in which first and second operations performed together is $\sum_{k_1=1}^{a_1} \sum_{k_2=1}^{a_2} C$. Now after it has been done the third operation can be performed in to say a_3 ways. Then we can write the number of ways in which first, second and third operations performed together is $\sum_{k_1=1}^{a_1} \sum_{k_2=1}^{a_2} \sum_{k_3=1}^{a_3} C$. Similarly the Vth operation can be performed in to say a_v ways and we can write the number of ways in which V operations performed together is $\sum_{k_1=1}^{a_1} \sum_{k_2=1}^{a_2} \sum_{k_3=1}^{a_3} \cdots \sum_{k_v=1}^{a_v} C$.

Example 14: Let 6 sided 4 dice is tossed. How many possible outcomes are there occurred.

Solution: Suppose we have 4 spaces to be filled with 6 digits 1, 2, 3, 4, 5, 6. The spaces can be designed as

$$k_1 \qquad\qquad k_1 \qquad\qquad k_1 \qquad\qquad k_1$$

Fig 1: Four space designation

We can fill the first space with one of the six digits (as the first die has 6 sides) and so in 6 ways i.e., $\sum_{k_1=1}^{6} C = 6 \times C = 6$ ways. After it is filled the second space can be filled with one of the six digits (as the second die has 6 sides) and so in 6 ways. Thus the two spaces can be filled together in $\sum_{k_1=1}^{6} \sum_{k_2=1}^{6} C$ ways i.e., 36 ways. After it is filled the third space can be filled with one of the six digits (as the third die has 6 sides) and so in 6 ways.

Thus the three spaces can be filled together in $\sum_{k_1=1}^{6} \sum_{k_2=1}^{6} \sum_{k_3=1}^{6} C$ ways i.e., 216 ways. After it is filled the fourth space can be filled with one of the six digits (as the fourth die has 6 sides) and so in 6 ways. Thus the four spaces can be filled together in $\sum_{k_1=1}^{6} \sum_{k_2=1}^{6} \sum_{k_3=1}^{6} \sum_{k_4=1}^{6} C$ ways i.e., 1296 ways.

5. Summation method III

Let an operation can be performed in $k_1{}^{th}$ way; $k_1 \in K_1$ and after it is performed in any one of these ways a second operation can be performed in $k_2{}^{th}$ way; $k_2 \in K_2$ and $\Phi : K_1 \to K_2$ and after it is performed in any one of these ways a third operation can be performed in $k_3{}^{th}$ way; $k_3 \in K_3$ and $\Phi : K_2 \to K_3$ and so on for V operations, then the V operations can be performed together in

$$\sum_{k_1 \in K_1} \sum_{k_2 \in K_2} \sum_{k_3 \in K_3} \cdots \sum_{k_V \in K_V} C = C_V \qquad\qquad \text{——————— (14)}$$

ways, where C is a constant quantity taking unit value and say

$$K_v = (K_{v.1}, K_{v.2}, K_{v.3}, \ldots, K_{v.g}, \ldots, K_{v.h})$$

C is the component number contained in the assembly A_V if the events contained in the assembly are of one component.

Proof: Let we have V operations to be performed. Since the first operation can be performed in $k_1{}^{th}$ way, $k_1 \in K_1$ we can write the number of ways in which first operation performed is

$$\sum_{k_1 \in K_1} C = C + C + C + \ldots\ldots + C \qquad\qquad \text{——————— (15)}$$

terms equal to the number of components of the assembly $K_1 = (K_{11})$ i.e., equal to $C_1 = C_{11}$

Thus (15) can be written as

$$\sum_{k_1 \in K_1} C = C_{11} = C_1 \qquad\qquad \text{——————— (16)}$$

After it has been done the second operation can be performed in $k_2{}^{th}$ way, $k_2 \in K_2$ and $\Phi : K_1 \to K_2$. Thus we can write the number of ways in which first and second operations performed together is

$$\sum_{k_1 \in K_1} \sum_{k_2 \in K_2} C$$
$$= \sum_{k_{21} \in K_{21}} C + \sum_{k_{22} \in K_{22}} C + \sum_{k_{23} \in K_{23}} C + \ldots + \sum_{k_{2C_1} \in K_{2C_1}} C \text{ ——— (17)}$$

terms equal to the number of components of the assembly $K_1 = (K_{11})$ i.e., equal to $C_1 = C_{11}$

Again the terms of the right side of (17) can be expanded as $C_{21}, C_{22}, C_{23},$ ..., C_{2C_1} terms respectively. Thus (17) can be written as

$$\sum_{k_1 \in K_1} \sum_{k_2 \in K_2} C = C_{21} + C_{22} + C_{23} + ... + C_{2C_1} = C_2 \quad \text{————————(18)}$$

After it has been done the third operation can be performed in $k_3{}^{th}$ way, $k_3 \in K_3$ and $\Phi : K_2 \to K_3$. Thus we can write the number of ways in which first, second and third operations performed together is

$$\sum_{k_1 \in K_1} \sum_{k_2 \in K_2} \sum_{k_3 \in K_3} C = \sum_{k_{21} \in K_{21}} \sum_{k_3 \in K_3} C + \sum_{k_{22} \in K_{22}} \sum_{k_3 \in K_3} C + \sum_{k_{23} \in K_{23}} \sum_{k_3 \in K_3} C + ... + \sum_{k_{2C_1} \in K_{2C_1}} \sum_{k_3 \in K_3} C \quad \text{————————(19)}$$

terms equal to the number of components of the assembly $K_1 = (K_{11})$ i.e., equal to $C_1 = C_{11}$

Again the terms of the right side of (19) can be expanded as (17). Thus (19) can be written as

$$\sum_{k_1 \in K_1} \sum_{k_2 \in K_2} \sum_{k_3 \in K_3} C$$

$$= (\sum_{k_{31} \in K_{31}} C + \sum_{k_{32} \in K_{32}} C + \sum_{k_{33} \in K_{33}} C + ... + \sum_{k_{3C_{21}} \in K_{3C_{21}}} C +)$$

$$+ (\sum_{k_{3(C_{21}+1)} \in K_{3(C_{21}+1)}} C + \sum_{k_{3(C_{21}+2)} \in K_{3(C_{21}+2)}} C + ...$$

$$+ \sum_{k_{3(C_{21}+C_{22})} \in K_{3(C_{21}+C_{22})}} C) + ...$$

$$+ (\sum_{k_{3(C_{21}+C_{22}+...+C_{2(C_1-1)}+1)} \in K_{3(C_{21}+C_{22}+...+C_{2(C_1-1)}+1)}} C$$

$$+ \sum_{k_{3(C_{21}+C_{22}+...+C_{2(C_1-1)}+2)} \in K_{3(C_{21}+C_{22}+...+C_{2(C_1-1)}+2)}} C + ...$$

$$+ \sum_{k_{3(C_{21}+C_{22}+...+C_{2C_1})} \in K_{3(C_{21}+C_{22}+...+C_{2C_1})}} C)$$

$$= (C_{31} + C_{32} + C_{33} + ... + C_{3C_{21}}) + (C_{3(C_{21}+1)} + C_{3(C_{21}+2)} + ...$$

$$+ C_{3(C_{21}+C_{22})}) + ... + (C_{3(C_{21}+C_{22}+...+C_{2(C_1-1)}+1)}$$

$$+ C_{3(C_{21}+C_{22}+...+C_{2(C_1-1)}+2)} + ... + C_{3(C_{21}+C_{22}+...+C_{2C_1})})$$

$$= C_{31} + C_{32} + C_{33} + ... + C_{3(C_{21}+C_{22}+...+C_{2C_1})}$$

$$= C_{31} + C_{32} + C_{33} + ... + C_{3C_2}$$

$$= C_3 \quad \text{————————————— (20)}$$

Proceeding similarly we get for V operations, the V operations can be performed together in

$$\sum_{k_1 \in K_1} \sum_{k_2 \in K_2} \sum_{k_3 \in K_3} \cdots \sum_{k_V \in K_V} C = C_V \text{ ways. Hence the proof.}$$

It is why to use the language like " ... performed in $k_V{}^{th}$ way " instead of " ... performed in k_V ways " as we have sometimes the assembly K_V need not to have from l_V through h_V integers. There may be some missing integers. We correspond such questions in the theorems later.

Example 15: Let there are 4 operations to be performed. The first operation can be performed in k_1^{th} way where $k_1 \in K_1$ and $K_1 = (1, 2, 3)$. After it is performed the second operation can be performed in k_2^{th} way where $k_2 \in K_2$ and $K_2 = (1, 2, 3, ..., k_1 + 1)$. After it is performed the third operation can be performed in k_3^{th} way where $k_3 \in K_3$ and $K_3 = (1, 2, 3, ..., k_2 + 1)$. And after it is performed the fourth operation can be performed in k_4^{th} way where $k_4 \in K_4$ and $K_4 = (1, 2, 3, ..., k_3 + 1)$. Find how many ways the 4-operations can be performed together in ?

Solution: We have given $K_1 = K_{11} = (1, 2, 3)$.

Thus integers contained in K_{11} i.e., the first operation can be performed in $\sum_{k_1 \in K_1} C = C_{11} = 3 = C_1$ ways.

Again $K_2 = (K_{21}, K_{22}, K_{23})$ where, $K_{2.1} = (1, 2)$
$$K_{2.2} = (1, 2, 3)$$
$$K_{2.3} = (1, 2, 3, 4)$$

Thus first and second operations can be performed together in $\sum_{k_1 \in K_1} \sum_{k_2 \in K_2} C = C_{21} + C_{22} + C_{23} = 2+3+4 = 9 = C_2$ ways.

Again $K_3 = (K_{31}, K_{32}, K_{33}, K_{34}, K_{35}, K_{36}, K_{37}, K_{38}, K_{39})$

where, $K_{3.1} = (1, 2)$ $\qquad K_{3.6} = (1, 2)$
$K_{3.2} = (1, 2, 3)$ $\qquad K_{3.7} = (1, 2, 3)$
$K_{3.3} = (1, 2)$ $\qquad K_{3.8} = (1, 2, 3, 4)$
$K_{3.4} = (1, 2, 3)$ $\qquad K_{3.9} = (1, 2, 3, 4, 5)$
$K_{3.5} = (1, 2, 3, 4)$

Thus first, second and third operations can be performed together in
$$\sum_{k_1 \in K_1} \sum_{k_2 \in K_2} \sum_{k_3 \in K_3} C$$
$$= C_{3.1} + C_{3.2} + C_{3.3} + C_{3.4} + C_{3.5} + C_{3.6} + C_{3.7} + C_{3.8} + C_{3.9}$$
$$= 2+3+2+3+4+2+3+4+5 = 28 \quad \text{ways.}$$

Again $K_4 = (K_{4.1}, K_{4.2}, K_{4.3}, K_{4.4}, K_{4.5}, K_{4.6}, K_{4.7}, K_{4.8}, K_{4.9}, K_{4.10}, K_{4.11}, K_{4.12}, K_{4.13}, K_{4.14}, K_{4.15}, K_{4.16}, K_{4.17}, K_{4.18}, K_{4.19}, K_{4.20}, K_{4.21}, K_{4.22}, K_{4.23}, K_{4.24}, K_{4.25}, K_{4.26}, K_{4.27}, K_{4.28})$

where, $K_{4.1} = (1, 2)$ $\qquad K_{4.15} = (1, 2)$
$K_{4.2} = (1, 2, 3)$ $\qquad K_{4.16} = (1, 2, 3)$
$K_{4.3} = (1, 2)$ $\qquad K_{4.17} = (1, 2)$

$K_{4.4} = (1, 2, 3)$ $K_{4.18} = (1, 2, 3)$

$K_{4.5} = (1, 2, 3, 4)$ $K_{4.19} = (1, 2, 3, 4)$

$K_{4.6} = (1, 2)$ $K_{4.20} = (1, 2)$

$K_{4.7} = (1, 2, 3)$ $K_{4.21} = (1, 2, 3)$

$K_{4.8} = (1, 2)$ $K_{4.22} = (1, 2, 3, 4)$

$K_{4.9} = (1, 2, 3)$ $K_{4.23} = (1, 2, 3, 4, 5)$

$K_{4.10} = (1, 2, 3, 4)$ $K_{4.24} = (1, 2)$

$K_{4.11} = (1, 2)$ $K_{4.25} = (1, 2, 3)$

$K_{4.12} = (1, 2, 3)$ $K_{4.26} = (1, 2, 3, 4)$

$K_{4.13} = (1, 2, 3, 4)$ $K_{4.27} = (1, 2, 3, 4, 5)$

$K_{4.14} = (1, 2, 3, 4, 5)$ $K_{4.28} = (1, 2, 3, 4, 5, 6)$

Thus first, second, third and fourth operations can be performed together in

$$\sum_{k_1 \in K_1} \sum_{k_2 \in K_2} \sum_{k_3 \in K_3} \sum_{k_4 \in K_4} C = C_{4.1} + C_{4.2} + C_{4.3} + C_{4.4} + C_{4.5} + C_{4.6}$$
$$+ C_{4.7} + C_{4.8} + C_{4.9} + C_{4.10} + C_{4.11} + C_{4.12} + C_{4.13} + C_{4.14} + C_{4.15} + C_{4.16}$$
$$+ C_{4.17} + C_{4.18} + C_{4.19} + C_{4.20} + C_{4.21} + C_{4.22} + C_{4.23} + C_{4.24} + C_{4.25}$$
$$+ C_{4.26} + C_{4.27} + C_{4.28}$$
$$= 2 + 3 + 2 + 3 + 4 + 2 + 3 + 2 + 3 + 4 + 2 + 3 + 4 + 5 + 2 + 3$$
$$+ 2 + 3 + 4 + 2 + 3 + 4 + 5 + 2 + 3 + 4 + 5 + 6$$
$$= 90 \text{ ways.}$$

Example 16: Consider the example 15 and find $\Phi : K_1 \to K_2$ and $\Phi : K_2 \to K_3$

Solution: The one-one mapping of K_1 onto K_2 is as follows:

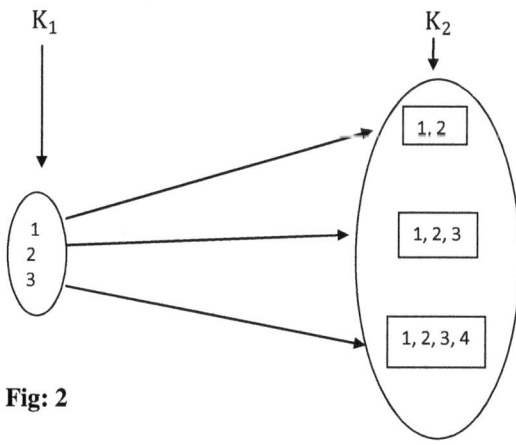

Fig: 2

18

The one-one mapping of K_2 onto K_3 is as follows:

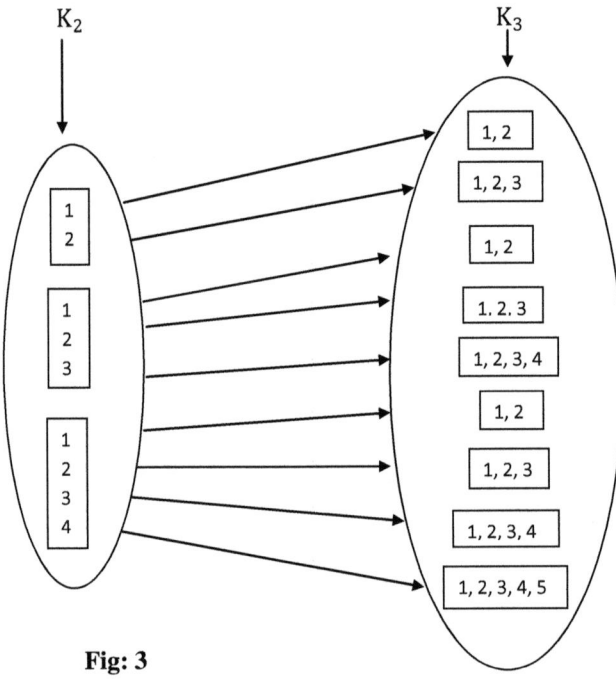

Fig: 3

6. Biswas triangle

The triangle constitutes the numbers

$$C\binom{R}{V}, \ C\binom{R-1}{V}, \ C\binom{R-2}{V}, \ ..., \ C\binom{V}{V}$$
$$C\binom{R-1}{V}, \ C\binom{R-2}{V}, \ ..., \ C\binom{V}{V}$$
$$C\binom{R-2}{V}, \ ..., \ C\binom{V}{V}$$
$$\vdots$$
$$C\binom{V}{V}$$

are to be called Biswas triangle. It is denoted by $B\Delta C\binom{R}{V}$ and reads

Biswas triangle topping $C\binom{R}{V}$, i.e.,

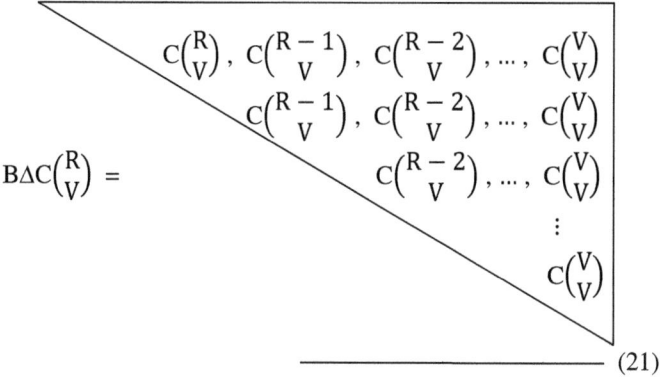

$$B\Delta C\binom{R}{V} = \begin{array}{l} C\binom{R}{V},\ C\binom{R-1}{V},\ C\binom{R-2}{V},\ ...,\ C\binom{V}{V} \\ C\binom{R-1}{V},\ C\binom{R-2}{V},\ ...,\ C\binom{V}{V} \\ C\binom{R-2}{V},\ ...,\ C\binom{V}{V} \\ \vdots \\ C\binom{V}{V} \end{array}$$

——————————————— (21)

The sum of numbers of Biswas triangle topping $C\binom{R}{V}$ is denoted by $SB\Delta C\binom{R}{V}$ i.e.,

$$SB\Delta C\binom{R}{V} = C\binom{R}{V} + 2C\binom{R-1}{V} + 3C\binom{R-2}{V} + ... + (R-V+1)C\binom{V}{V}$$

——————————————— (22)

Example 17: Write down the Biswas triangle topping $C\binom{5}{2}$

Solution: The desired Biswas triangle is

$$C\binom{5}{2} = \begin{array}{l} C\binom{5}{2},\ C\binom{4}{2},\ C\binom{3}{2},\ C\binom{2}{2} \\ C\binom{4}{2},\ C\binom{3}{2},\ C\binom{2}{2} \\ C\binom{3}{2},\ C\binom{2}{2} \\ C\binom{2}{2} \end{array}$$

The following theorem states the sum of numbers of Biswas triangle topping $C\binom{R}{V}$.

Theorem 1: The sum of numbers of Biswas Triangle topping $C\binom{R}{V}$, denoted by $SB\Delta C\binom{R}{V}$ is

$$SB\Delta C\binom{R}{V} = C\binom{R+2}{V+2} \qquad\qquad \text{(23)}$$

Proof: We get

$$SB\Delta C\binom{R}{V} = C\binom{R}{V} + 2C\binom{R-1}{V} + 3C\binom{R-2}{V} + ... + (R-V+1)C\binom{V}{V}$$

$$= \frac{1}{V!} [\ 1.R(R-1)(R-2).....(R-V+1)$$

$$+ 2.(R-1)(R-2)(R-3).... (R-V)$$

$$+ 3.(R-2)(R-3)(R-4)....(R-V-1) + ...$$

$$+ (R-V-1).(V+2)(V+1)(V)....3$$

$$+ (R-V).(V+1)(V)(V-1).....2$$

$$+ (R-V+1). (V)(V-1)(V-2).....1\ ]$$

Writing the series from last term we get

$$SB\Delta C\binom{R}{V} = \frac{1}{V!} [\ (R-V+1).1.2.3......V + (R-V).2.3.4.......(V+1)$$

$$+ (R-V-1).3.4.5.......(V+2) + ...$$

$$+ 3.(R-V-1)(R-V)(R-V+1).......(R-2)$$

$$+ 2.(R-V)(R-V+1)(R-V+2).......(R-1)$$

$$+ 1.(R-V+1)(R-V+2)(R-V+3).......R\ ]$$

$$= \frac{1}{V!} [\ (R-V+1)a_1 + (R-V)a_2 + (R-V-1)a_3 + ... + 3.a_{(R-V-1)}$$

$$+ 2.a_{(R-V)} + 1.a_{(R-V+1)}\ ]$$

where, $a_1 = 1.2.3.....V$

$\qquad\qquad a_2 = 2.3.4.....(V+1)$

$\qquad\qquad a_3 = 3.4.5.....(V+2)$

$\qquad\qquad \vdots$

$\qquad\qquad a_{(R-V-1)} = (R-V-1)(R-V)(R-V+1).....(R-2)$

$\qquad\qquad a_{(R-V)} = (R-V)(R-V+1)(R-V+2).....(R-1)$

$\qquad\qquad a_{(R-V+1)} = (R-V+1)(R-V+2)(R-V+3).....R$

Now using the method of difference we get the sum

$$SB\Delta C\binom{R}{V} = [(R-V+1)(b_1-b_0) + (R-V)(b_2-b_1) + (R-V-1)(b_3-b_2)$$

$$+ ...... + 3.(b_{(R-V-1)} - b_{(R-V-2)}) + 2.(b_{(R-V)} - b_{(R-V-1)})$$

$$+ 1.(b_{(R-V+1)} - b_{(R-V)})]$$

$$= [-(R-V+1)b_0 + b_1 + b_2 + \dots\dots + b_{(R-V-1)} + b_{(R-V)}$$

$$+ b_{(R-V+1)}]$$

where, $b_0 = 0$

$b_1 = 1.2.3\dots V(V+1)/(V+1)$

$b_2 = 2.3.4\dots(V+1)(V+2)/(V+1)$

$\vdots$

$b_{(R-V-1)} = (R-V-1)(R-V)(R-V+1)\dots(R-2)(R-1)/(V+1)$

$b_{(R-V)} = (R-V)(R-V+1)(R-V+2)\dots(R-1)R/(V+1)$

$b_{(R-V+1)} = (R-V+1)(R-V+2)(R-V+3)\dots R(R+1)/(V+1)$

Thus the sum is

$$SB\Delta C\binom{R}{V} = \frac{1}{(V+1)V!}[1.2.3\dots(V+1) + 2.3.4\dots(V+2)$$

$$+ 3.4.5\dots(V+3) + \dots$$

$$+ (R-V-1)(R-V)(R-V+1)\dots(R-1)$$

$$+ (R-V)(R-V+1)(R-V+2)\dots R$$

$$+ (R-V+1)(R-V+2)(R-V+3)\dots(R+1)]$$

Again using the method of difference we get

$$SB\Delta C\binom{R}{V} = \frac{1}{(V+1)!}[b'_{(R-V+1)} - b'_0]$$

where, $b'_{(R-V+1)} = (R-V+1)(R-V+2)(R-V+3)\dots(R+1)(R+2)/(V+2)$

$b'_0 = 0$

Thus the sum is

$$SB\Delta C\binom{R}{V} = \frac{(R-V+1)(R-V+2)(R-V+3)\dots(R+1)(R+2)}{(V+2)(V+1)!}$$

$$= \frac{(R+2)(R+1)(R)\dots(R-V+3)(R-V+2)(R-V+1)}{(V+2)!}$$

$$= \frac{(R+2)!}{(V+2)!(R-V)!}$$

$$= C\binom{R+2}{V+2}$$

Hence the proof.

Example 18: Find the sums of numbers of B triangle topping (i) $C\binom{5}{2}$ and (ii) $C\binom{6}{3}$

Solution: (i) We get from equation (23)

$$\text{SBΔC}\binom{5}{2} = C\binom{5+2}{2+2} = C\binom{7}{4} = 35$$

(ii) We get from equation (23)

$$\text{SBΔC}\binom{6}{3} = C\binom{6+2}{3+2} = C\binom{8}{5} = 56.$$

7. Summation method IV

Let $\quad l_1 = 1 \qquad\qquad h_1 = (R-V+1)$

$\qquad\quad l_2 = 1 \qquad\qquad h_2 = (R-V+2-k_1)$

$\qquad\quad l_3 = 1 \qquad\qquad h_3 = (R-V+3-k_1-k_2)$

$\qquad\quad \vdots$

$\qquad\quad l_V = 1 \qquad\qquad h_V = (R-k_1-k_2- \ldots\ldots -k_{V-1})$

then the sum

$$\sum_{k_1=l_1}^{h_1} \sum_{k_2=l_2}^{h_2} \sum_{k_3=l_3}^{h_3} \ldots \sum_{k_V=l_V}^{h_V} C = C\binom{R}{V} \quad\quad\text{——————} \quad (24)$$

Proof: Let the sum

$$\sum_{k_V=1}^{(R-V+1)} C + \sum_{k_V=1}^{(R-V)} C + \sum_{k_V=1}^{(R-V-1)} C + \ldots\ldots + \sum_{k_V=1}^{1} C$$

$$= \sum_{k_{V-1}=1}^{(R-V+1)} \sum_{k_V=1}^{(R-V+2)-k_{V-1}} C$$

$$= (R-V+1) + (R-V) + (R-V-1) + \ldots\ldots + 1$$

$$= \frac{(R-V+1)\,(R-V+2)}{2}$$

$$= C\binom{R-V+2}{2} \quad\quad\text{——————} \quad (25)$$

And $\quad \sum_{k_{V-1}=1}^{(R-V+1)} \sum_{k_V=1}^{(R-V+2)-k_{V-1}} C + \sum_{k_{V-1}=1}^{(R-V)} \sum_{k_V=1}^{(R-V+1)-k_{V-1}} C + \ldots.. +$

$\sum_{k_{V-1}=1}^{1} \sum_{k_V=1}^{2-k_{V-1}} C$

$$= \sum_{k_{V-2}=1}^{(R-V+1)} \sum_{k_{V-1}=1}^{(R-V+2)-k_{V-2}} \sum_{k_V=1}^{(R-V+3)-k_{V-2}-k_{V-1}} C$$

$$= C\binom{R-V+2}{2} + C\binom{R-V+1}{2} + C\binom{R-V}{2} + \ldots\ldots + C\binom{2}{2} \quad\text{——} \quad (26)$$

And $\quad \sum_{k_{V-1}=1}^{(R-V)} \sum_{k_V=1}^{(R-V+1)-k_{V-1}} C + \sum_{k_{V-1}=1}^{(R-V-1)} \sum_{k_V=1}^{(R-V)-k_{V-1}} C + \ldots.. +$

$\sum_{k_{V-1}=1}^{1} \sum_{k_V=1}^{2-k_{V-1}} C$

$$= \sum_{k_{V-2}=1}^{(R-V)} \sum_{k_{V-1}=1}^{(R-V+1)-k_{V-2}} \sum_{k_V=1}^{(R-V+2)-k_{V-2}-k_{V-1}} C$$

$$= C\binom{R-V+1}{2} + C\binom{R-V}{2} + C\binom{R-V-1}{2} + \ldots.. + C\binom{2}{2} \quad\text{——} \quad (27)$$

23

Similarly $\sum_{k_{V-2}=1}^{(R-V-1)} \sum_{k_{V-1}=1}^{(R-V)-k_{V-2}} \sum_{k_V=1}^{(R-V+1)-k_{V-2}-k_{V-1}} C$

$= C\binom{R-V}{2} + C\binom{R-V-1}{2} + C\binom{R-V-2}{2} + \ldots\ldots + C\binom{2}{2}$ —— (28)

And so on $\sum_{k_{V-2}=1}^{1} \sum_{k_{V-1}=1}^{2-k_{V-2}} \sum_{k_V=1}^{3-k_{V-2}-k_{V-1}} C = C\binom{2}{2}$ ———— (29)

Now (26), (27), (28) and (29) make

$\sum_{k_{V-2}=1}^{(R-V+1)} \sum_{k_{V-1}=1}^{(R-V+2)-k_{V-2}} \sum_{k_V=1}^{(R-V+3)-k_{V-2}-k_{V-1}} C$

$\quad + \sum_{k_{V-2}=1}^{(R-V)} \sum_{k_{V-1}=1}^{(R-V+1)-k_{V-2}} \sum_{k_V=1}^{(R-V+2)-k_{V-2}-k_{V-1}} C$

$\quad + \sum_{k_{V-2}=1}^{(R-V-1)} \sum_{k_{V-1}=1}^{(R-V)-k_{V-2}} \sum_{k_V=1}^{(R-V+1)-k_{V-2}-k_{V-1}} C + \ldots..$

$\quad + \sum_{k_{V-2}=1}^{1} \sum_{k_{V-1}=1}^{2-k_{V-2}} \sum_{k_V=1}^{3-k_{V-2}-k_{V-1}} C$

$= C\binom{R-V+2}{2} + 2C\binom{R-V+1}{2} + 3C\binom{R-V}{2} + \ldots + (R-V+1)C\binom{2}{2}$

i.e.,

$\sum_{k_{V-3}=1}^{(R-V+1)} \sum_{k_{V-2}=1}^{(R-V+2)-k_{V-3}} \sum_{k_{V-1}=1}^{(R-V+3)-k_{V-3}-k_{V-2}} \sum_{k_V=1}^{(R-V+4)-k_{V-3}-k_{V-2}-k_{V-1}} C$

$\quad = C\binom{R-V+4}{4}$ ———————— (30)

Proceeding these ways (25) to (30) we get

$\sum_{k_{V-5}=1}^{(R-V+1)} \sum_{k_{V-4}=1}^{(R-V+2)-k_{V-5}} \sum_{k_{V-3}=1}^{(R-V+3)-k_{V-5}-k_{V-4}} \cdots \sum_{k_V=1}^{(R-V+6)-k_{V-5}-k_{V-4}\cdots-k_{V-1}} C$

$\quad = C\binom{R-V+6}{6}$

and

$\sum_{k_{V-7}=1}^{(R-V+1)} \sum_{k_{V-6}=1}^{(R-V+2)-k_{V-7}} \sum_{k_{V-5}=1}^{(R-V+3)-k_{V-7}-k_{V-6}} \cdots \sum_{k_V=1}^{(R-V+8)-k_{V-7}-k_{V-6}\cdots-k_{V-1}} C$

$\quad = C\binom{R-V+8}{8}$

and so on

$\sum_{k_1=1}^{(R-V+1)} \sum_{k_2=1}^{(R-V+2)-k_1} \sum_{k_3=1}^{(R-V+3)-k_1-k_2} \cdots \sum_{k_V=1}^{R-k_1-k_2-\cdots-k_{V-1}} C$

$\quad = C\binom{R-V+V}{V} = C\binom{R}{V}$

Hence the proof.

Example 19: Let there are 6 operations. First operation can be performed in 4 ways and after it is done second operation can be performed in $4-k_1$ ways where $k_1 = 1, 2, 3, 4$ and after it is done third operation can be performed in $6-k_1-k_2$ ways where $k_2 = 1, 2, 3,\ldots, 5-k_1$ and so on 6th operation can be performed in $9-k_1-k_2-\ldots-k_5$ ways where $k_5 = 1, 2, 3,\ldots,$

$8 - k_1 - k_2 - .... - k_4$. Find the number of ways the 6 operations can be performed together in.

Solution: The number of ways in which the 6 operations can be performed together in is

$$\sum_{k_1=1}^{4} \sum_{k_2=1}^{5-k_1} \sum_{k_3=1}^{6-k_1-k_2} \sum_{k_4=1}^{7-k_1-k_2-k_3} \sum_{k_5=1}^{8-k_1-k_2-k_3-k_4} \sum_{k_6=1}^{9-k_1-k_2-k_3-k_4-k_5} C$$

$$= C\binom{9}{6} = 84.$$

8. Conclusions

The topic widely used to explain theorems of partitions, factorizations, combinations, permutations, formations, homogenations etc. It also used in statistical data analysis.

References

1. Deapon Biswas, Paper 2, Assemblies, Bystematics My Classic, March 2010 Self published, Chittagong, February 2016 Monon Prokashon, Chittagong, Bystematics Vol. I, My Classic, March 2018 Scholar's Press EU, ISBN: 987- 620-2-30664-5.

2. Deapon Biswas, Paper 3, B relations, Bystematics My Classic, March 2010 Self published, Chittagong, February 2016 Monon Prokashon, Chittagong, Bystematics Vol. I, My Classic, March 2018 Scholar's Press EU, ISBN: 987- 620-2-30664-5.

3. Deapon Biswas, Paper 4, B space, Bystematics My Classic, March 2010 Self published, Chittagong, February 2016 Monon Prokashon, Chittagong, Bystematics Vol. I, My Classic, March 2018 Scholar's Press EU, ISBN: 987- 620-2-30664-5.

4. Deapon Biswas, Paper 6, Summation methods Bystematics My Classic, 2010 Self published, Chittagong, 2016 Monon Prokashon, Chittagong, Bystematics Vol. I, My Classic, 2018 Scholar's Press EU, ISBN: 987- 620-2-30664-5.

5. Carl B. Allendoerfer & Cletus O. Oakley. Principles of Mathematics.

6. A. K. S. M. Nawab Ali, Advanced Algebra.

YOUR KNOWLEDGE HAS VALUE

- We will publish your bachelor's and
  master's thesis, essays and papers

- Your own eBook and book -
  sold worldwide in all relevant shops

- Earn money with each sale

Upload your text at www.GRIN.com
and publish for free